Phony Phorestry

A Forestry Insider Calls Out The Forestry Profession
First Edition—November 2015

By Jim Birkemeier—Timber Grower

Phony Phorestry

An Insider Calls Out the Forestry Profession:

Government Foresters make Forestry a Welfare Program

University Foresters Legitimize Industrial Logging as Sustainable Forestry

Private Forestry Companies take advantage of the system to sustain their income

Retired government foresters who go into business with their insider information 'double dip'

They all tell landowners "you can't make any money growing trees—so we will subsidize your forest"

Forest owners are the bottom of the food chain in the timber industry—in their turn they are lead to the slaughter

Forest management should be a profitable business that pays its own way and supports the local community

After 3 years in business, I had to get out
I could not accept what I saw all around
Business as usual was bad for the forest

High — Grade Logging must be stopped

'Taking just the Best and Leaving the Rest'

**Our family re-learned the forestry business
from the timber growers point of view**

No compromise — we just do what is right

Here is an alternative to Phony Phorestry

Jim Birkemeier—Timber Grower

(professional forester—Bachelor Degree of Forest Science UW Madison-76)

Phony Phorestry

Contents:

Jim Birkemeier—forester and timber grower
November 2015

Forest owners face a new quagmire in the timber industry today, that makes things worse than ever before, to every landowner with trees today. The State Wisconsin has recently adapted the practice of feeding the big timber industry for the 'good of society.'

Retired DNR foresters go into business today as "consulting foresters" to help landowners manage and sell their trees as "Sustainable Forestry" in the Mandatory Harvests of the Wisconsin Managed Forest Law. They use their insider information and good guy reputation and former State Forester's Badge to fleece the innocent landowners for their own greed.

of

Many DNR foresters now plan their retirement for a decade or more, by writing and approving management plans to schedule harvests to carefully cash in on the cushy 10% commission for mandatory timber sales for the best forest resources of the landowners in the county, for their own personal profit. They likely feel entitled to get in on the 'gravy train' of foresters getting big commission checks for administering timber sales for landowners they just watched roll on by for their years as a public servant.

With their insider information and influence, their double dipping can now be multiplied over and over since no authorities care about trees or forests or landowners rights. Any reported scandal is ignored or covered up by State Foresters to protect their cronies and insure the system is working well when they retire.

Forest owners agree to enter the Managed Forest Law under the guise that it is Sustainable and Sound Forestry. The forestry profession suckers the landowners into selling their timber for about 10% of the cost of growing and managing timber crops, to feed the big fish of the timber industry while the foresters collect a professional salary with benefits. Sustainable Forestry is whatever the forester thinks will sustain their own salary.

Retired Wisconsin DNR Foresters go into business as "consulting foresters" and now use their insider information to take advantage of the entire timber marketing system for their personal greed. The forest owners and the forest resource and the taxpayers are the big losers.

Starting about 25 years ago, the "county forester" we all know and love – actually a DNR employee - began the now popular trend of taking early retirement, escaping the poorly run and frustrating government forestry programs.

4

Now receiving their pension and benefits, they use their nice guy trusted public servant personality and their insider information, to fleece the innocent and ignorant forest owners and the timber resource they formerly served as trusted government workers.

This practice has become so widespread today that it has changed the whole face of the timber industry in Wisconsin. The "timber buyer" who went out to contact landowners is rarely needed anymore. Now a big mill just has to open their mail and see all the forests set to harvest from the consulting foresters.

Using the Wisconsin Managed Forest Law, all embellished in the words and talk of Sustainable and Sound Forestry, foresters of all kinds entice landowners into property tax breaks only to lead them into forced timber harvests at market prices way below the costs of owning and growing timber crops. Professional foresters now have perfected the system to convince landowners to agree to subsidize the greedy global industrial timber industry at their own expense while supposedly feeling good about it.

Most "consulting foresters" are paid on a 10% commission for marking timber and soliciting bids for a landowner. This is a blatant conflict of interest that absolutely violates a professional forester's ethics and training. For the forestry profession to accept and protect such an extreme conflict of interest today demands immediate attention, full exposure, and reform.

Independent consulting foresters are now required to take DNR training and follow DNR rules to qualify to participate in the Managed Forest Law programs and other government funding systems. You have to play their games and follow their rules or be blocked out of the 'gravy train' of government funding.

Over the years I have blown the whistle to try to attract attention to this ongoing fraud. All attempts to expose the scandals have been covered up by the leadership of the DNR, to protect their buddies and their own hopes of cashing in on this forestry feeding frenzy when it is time for them to retire also.

This is all grown to the point where the timber market is flooded with timber sales prepared by consulting foresters. Sawmills and timber companies no longer need to employ a team of timber buyers to go out to contact landowners to find wood to harvest – they just have to open their mail and choose from the multitude of timber sales now on the market through the consulting foresters.

Forester after DNR forester I used to work with when I was younger have now retired and used their insider information, influence, nice guy reputation, to cash in on the easy money marking timber and soliciting bids for landowners. Recent retirees have spent their last decade or two planning for this time in their careers. They have written management plans for landowners and laid the groundwork for scheduling the mandatory and "sustainable" harvests for their friendly and trusting landowners for their scheduled time of early retirement.

Everything is masked in the idea that it is good and responsible and even mandatory that the forest owners now harvest their timber at prices that average about 10 percent of the costs of owning and managing timber crops.   In these operations the retired DNR foresters collect huge commission checks while still getting their pensions and benefits from the State.

These foresters use the fear of bad logging to get landowners to pay them for protection, but when the harvest starts, the landowner find they have been 'thrown to the wolves' and no one cares anywhere.

This is totally unfair competition to the self-employed consulting forestry companies who also offer the same services to landowners. But these guys and gals have to bow to the DNR to survive in the market.

Old guys who retire, become expert consultants, get new income streams while collecting their pension and benefits from the State also continue the same old traditional disastrous stream of industrial logging that is killing our local economy and block any new ideas from our young people who might hope of bringing about good positive change for the future. This is likely the most important factor in all this – retired government workers who become consultants continue the big corporation dominance of the markets that export our money and jobs for the greed of a few distant people who take advantage of the people and resources of the planet. Influence peddling in the forest goes on unseen today.

The forest management and timber harvesting system are becoming more inbred, self -controlled, one sided, greedy, secretive, protective, and shameful. Landowners are more dominated, held in the dark, degraded, controlled, and gain less and less in the industrial global forestry system.   Things are getting worse and worse by the day.

There is an alternative to all this. Promoting the use of locally grown and manufactured forest products would change all of this right away. It would rebuild our local economies and provide hundreds of thousands of good rewarding jobs, while directly reducing the demand to clear cut the remaining tropical rainforests and stop the illegal logging of the remnants of natural forest reserves.

Grown and Made in Wisconsin, Grown and Made in The USA Should be the focus of government forestry services, not just taking their share of the exporting of our best timber to foreign corporations.

There is plenty of value to finished wood products to support the local community, if the wood is processed here, not on the other side of the planet by the cheapest labor working in the worst conditions with no regard to protection of the environment.

Foresters and Educators must tell the whole picture of growing, manufacturing, shipping, and using wood products so consumers can really decide what products are best to buy.

Phony Phorestry   An insider's call out of the forestry profession

If a man blows a whistle in the forest, and nobody cares...

Forests cover a third of our nation, nearly half of Wisconsin, yet timber is our most unknown and mismanaged and neglected of all our major agricultural crops. Compared to dairy, corn, beef, beans, alfalfa, etc.... knowledge and management of our trees and timber crops are a disastrous failure of our Universities and government forestry programs—to our society today in 2015.

This Beef Farmer works hard to optimize the production of his alfalfa, soy beans, corn, and beef herd. He knows everything about these crops. The forests in the background and across the region are neglected and mismanaged, producing about one quarter of their potential growth and economic return. Farmers know next to nothing about this huge and valuable agricultural crop that they all own. Foresters tell landowners that "only a professional forester" can manage trees.

25% gets an F in College and High School and Grade School....

Our forests produce about 25% of their potential in both annual volume growth and quality and economic return to Wisconsin. Market prices for trees in the timber industry pays the forest owner about 10% of the costs of owning and growing timber, so way less than 1 percent of forest owners manage their forests as any kind of profitable business. No other profession on the planet could survive on such disastrous figures, yet foresters act like everything is going along just fine and say forestry today is Sound and is really 'Sustainable' – even "Certified".

Foresters work together to keep landowners ignorant and disconnected from their timber . Foresters tell forest owners that 'only a professional is qualified to manage timber'

'don't even try to manage your own forest.'

'it's too dangerous for a landowner to use a chain saw to fell trees'.

Foresters herd the people into obedient docile flocks – fatten them up for a decade or two.   Then in turn, the foresters schedule the landowners and their trees for the industrial slaughter. The foresters take their professional salary and commissions and benefits and.... then abandon those timber growers so they can go fatten up some others for the next fleecing.

No professional forester could live on the value of the advice they provide to forest owners.

There is an alternative to this Phony Phoresty system.

I teach safe, common sense, practical forest management to other forest owners to share how we make hundreds of high value wood products from our dead and dying trees here at Timbergreen Farm—Spring Green Wisconsin

Wisconsin's Governor Walker boasts each yeat that our state is #1 in timber, but what he doesn't know is the trend that our state has lost over a half million jobs in the timber industry over the past few decades. Wisconsin's forest products output is down 50% also in the same time, from a high of $36 billion/year to the current less than $20 Billion/yr.   So while we may be still the #1 timber producing state in the USA, he is missing the point – something is very wrong here, but the forestry advisors always tell him everything is good and sustainable and now certified!   **Phoolish**

A neighbor in the valley is selling trees to a local logger. Some trees were marked, others show no paint. It appears the loggers cut anything they could to make money. The landowner may get about 1% of the value of the wood products – maybe 10% of their cost to own land and grow trees.   The damaged forest may regrow to produce another poor quality harvest in 20 years.   All the professional foresters say that everything in Wisconsin is Sustainable and Industrial Logging is Sound Forestry. Most Foresters call anything "Sustainable" as long as their own paycheck is sustained.

We could do so much better. We Should Do So Much Better.

There is a choice anyone can make today—a better way.

The past few decades our timber industry has been leaving and downsizing and outsourcing due to globalization – Wisconsin's 16 million acres of forests are today mostly neglected and abused – our urban trees are mostly chipped up into mountains of low value mulch or thrown in the landfill - and most of the wood products we buy in our stores now are cheap stuff imported from other countries far away.

Our own government and the people of Wisconsin do not support the state's timber industry, preferring to buy cheap imported substitutes that degrade the planet's environment in a multitude of ways. Cheap stuff "On Sale" at the neighborhood Big Box Store comes at a very high price to our planet and society. "Free Shipping" at the online Mega Store also costs us dearly in many ways. You get what you pay for.

We Could Choose to Use Good Local Wood

Here is a huge economic opportunity that we could take advantage of immediately. If we simply chose to buy and use wood products grown and manufactured in Wisconsin, within one year………
we could add Billions of Dollars and a hundred thousand jobs to our States' economy.

I have a BS Degree in Forest Science from the University of Wisconsin -Madison 1976 and have worked as a "professional forester" with small private woodland owners for 4 decades. As a timber grower for 47 years, our family tried everything good forest owners are supposed to do according to the professional foresters and government programs – and gradually saw that it was all Phony Phorestry. We Found A Better Way— **Do Just the Opposite** of what Industry, Government, and the Professionals tell us is sound forestry.

Our family wood business here at Spring Green Wisconsin has now learned to earn thousands of dollars per dead (really mature) tree, export high value finished products while importing money, keeping the income in the local economy, and can create a good job for every 10 acres of forest growth. We are in full control of our forest and our land—no professional experts needed. (No Professional forester even KNOWS what we do!!)

So the full potential for Wisconsin would be to produce $60 Billion per year, employing over one million people. Even if we take advantage of just a small percentage of the opportunity, it would create a large growth in our economy. It is well past time to break up this logjam.

While I live and work in Wisconsin, I have been around the world ten times this past decade, meeting with other forest owners and sharing with other land owners . Visitors from about 80 countries have visit us here at Timbergreen Farm to see what we do. Members of Parliament of Burma recently toured our business and enjoyed our Oak forests. (right)

The ideas here are totally universal in the global economy and timber industry. Wood is Wood

Trees are Trees   People are people.
We need to work together now to live together!

If a man blows a whistle in the forest, and nobody cares

does it make any sound??

does anybody care?   Do You?

Anywhere?

This is just a little blast by one little guy with a small forest in Wisconsin – to try to start the break-up of a huge historical industrial logjam. A million times more information and details are available and needed to get things moving in the right direction – if anyone cares somewhere?
Does anyone care to do what is right for our future?

The worst of the worst—from my experience

"Gang Rape in the Woods Goes Unnoticed_

I have had this posted on my website for years—no one seems to care.

"Two recent incidents made me so sick of foresters, I disavow any connection to the forestry profession.

Both started with retired DNR foresters who were now working for forest owners as consulting foresters. It seems that after many years of working the government system, some foresters want to cash in on some easy money in timber sales. They have an appearance of respect after working for the state, and everyone says they are 'real nice guys'. They play and talk the game of "Sound Forestry" to justify heavy timber harvests and they take a percentage of the sale income.

Both incidents had excellent written "forest management plans" in place – the language was perfect – it would be good for the woods to perform this thinning. Both forest owners had worked with DNR for years. They had done 'everything right' in planning to sell their timber (according to the "experts")

In both cases, the consulting foresters ignored their training & ethics and administered a greedy and destructive harvest. Quick profit for the forester and timber companies were their main concern. The values obtained for the forest owners were very low, even though excellent quality timber was taken. Heavy damage was done to the soils. Just the good timber was taken, leaving a highly degraded forest for the future. The management plans were not followed or respected. The good, high value trees were all taken.

The landowners were shocked by the damage and waste.

Both landowners appealed to the DNR after the harvests, and were not at all satisfied with their response. Both contacted me for my help.

Case One – Marquette County.   The landowner and I met with then State Forester Gene Francisco, Paul DeLong (current State Forester), Paul Pingrey, the consultant, and the procurement forester from the sawmill that bought the timber. The regional DNR forester, and two county DNR foresters observed the harvest on other visits. Francisco's verbal summary while observing the harvest was that this was just a 'typical timber harvest'. Nothing was done wrong. No action would be taken. The group concluded that the forest owner must have wanted to make as much money as possible and chose to have the heavy harvest. We asked other consulting foresters to assess the harvest and not one would help the forest owner, as they all knew they needed to keep on a good working relationship with the big sawmill.   The legal system was of no help. The individual forest owner is helpless in the timber industry.

The State Forester, Gene Francisco, told the forest owner and me that this looked like a 'typical timber harvest to him.' What's the problem? Paul DeLong stood by silent.

The forest owner got a fraction of the actual market value - the bid system was corrupt! Only a few bids were received for this 'best sale ever' in the county — and the big sawmill put in two bids, including one really low bid to make the other bid look acceptable. The old retired DNR forester and the Big Sawmill looked very suspicious!

When the ruts got too deep so the belly pan of the prehauler was dragging, the logger moved over and made a new set of ruts....   The damage was the worst I've ever seen.

The DNR obviously was more protective of the timber industry they work with every day and the landowner was blamed for wanting it.

Paul Pingrey did speak to me as we left— "You were a real Bull Dog for the landowner" All my efforts didn't do any good though - things just get worse and worse for the forest.

This red oak was growing 1/2" diameter each year
At Full Vigor!!

The owner did what the foresters told him was Sustainable Forest Management— a responsible and sound thinning.

He did everything a good forest owner is supposed to do.

The foresters scare the landowners into schools of 'sardines'

When it is their turn, they are given over to the big sharks in the industry.

An individual timber grower has no chance for fair prices.

The forest owner stands in what was the best oak forest in the county (right).
The landowner's son (left) looks over what remains of the forest that will some day be his to manage. The damage could never be repaired in his lifetime... Few families think about making money growing trees, they do it believing that they are doing the right thing. They often think that they can trust a professional forester and commonly get left with a disaster and no one to help. Even when a landowner get full market price and reasonable logging—they are only paid a very small fraction of what it costs to own and grow a forest. The foresters all collect a "professional salary".

Case Two – Iowa County. Again, a "respected" retired DNR forester, working as a consulting forester for the landowner - administered a devastating harvest in total violation of the forest management plan. Only the good big trees were taken, the forest owner got a fraction of the real market value, and heavy damage was done to the soil, roads, and forest. The landowner contacted DNR foresters Tom Hill (in the local office who wrote the management plan), Carol Nielsen, and Paul Pingrey at the Madison Office. The owner wrote many letters, none of which received any professional response. No one bothered to look at the forest and the damage – they "ignored her as an old lady in the retirement home who didn't know any better". My letters were ignored also. They all covered up the forest rape perpetrated by their friend and former co-worker. As far as I know, no action was ever taken. Last time I talked with the forest owner, she was so upset with the situation that she put the land up for sale.

These high value oak trees were growing with Full Vigor! The management plan was very specific - Leave the good big trees, and thin the stand by removing just some of the smaller trees. Only the biggest and best trees were actually harvested. The consulting forester provided a timber sale contract that gave permission to the logger to actually take the best unmarked trees—*but you'll have to pay for them if you get caught!!*

In both cases, retired DNR foresters working as consulting foresters 'raped' a forest and the forest owner – taking advantage of their innocent trust. The present DNR foresters worked to cover up the incidents and protect their friends. No one did a thing to stop it – or prevent it from happening again. This is getting worse—more foresters are doing it.

On a regular basis, other landowners call me to ask for help in dealing with foresters and the Managed Forest Law—usually too late to do anything. This is why I write this book.

The current compromised system of forestry

I'm a landowner, forester, logger, sawmiller, woodworker, and salesman. We developed our woodworking

business to earn enough money to pay a professional timber harvester a good wage for doing excellent work. My goal is to support good timber harvesting for the future of our planet. Harvesting is where we change the forest – current methods of industrial logging are simply 'taking' for short term greed. Everyone in the local community should benefit from the high value of wood products, not just a few rich people far away from the forest.

The Timber Industry is big business – it is reasonable to accept that they work to make their profit. My criticism comes when someone tries to represent taking trees in industrial logging and calling it sound and sustainable forestry. When a huge corporation takes the natural resources from a community for the profit of a few distant rich people, but leaves little benefit for the local people & a big mess for the future, I object!

There is a major traditional forest management system operating in our governments, industry, and universities that supports and justifies and protects industrial logging as good forestry. This satisfies people who only think about their own job. Everyone really knows deep down inside that things are not working as they should, but fear maintains the status quo. I see it all as Phony Phorestry.

Wisconsin's economy is getting about 25% of the potential value of our forest growth though the existing market system. Landowners do get a little money every decade or two, the industry still makes some products, most of our good logs and lumber are exported in low value form to the far east.

No other crop would be allowed to drag along at such a low level. While something is better than nothing, we should be getting much greater benefits for everyone—even the foresters and loggers— and the timber growers.

No one knows much about timber and few care – everything seems to be just fine

IN Wisconsin;

Most of our sawmills have closed in the past 25 years and half of the paper mill production has shut down.   Much of our furniture and flooring factories and veneer mills have shut down—due to globalization.

Landowners typically accept a system of allowing a logger to take what they can get every 15-20 years for a windfall payment that is "better than nothing".   Forest owners hope they get a good deal, rarely do, but no one really knows. No one is happy with this system, but there has been no alternative. (Today—there is a new choice)

For the forestry profession and our leaders to continue to support and protect this phony Phorestry system, and ignore & block all other alternatives is absurd. When everyone is talking jobs and economic growth, to ignore this opportunity in our local trees should not be tolerated.

A more active use of local wood would benefit everyone greatly, yet the professionals are all afraid to change and learn something new. They refuse to face reality today and fight to protect their phony phorestry.

As the timber industry fell apart the past 25 years here due to globalization, their pressure on the government to put timber on the market has dropped way off. But little has changed.

Government Forestry Programs benefit Industry, not the forest or forest owner

The Wisconsin Managed Forest Law entices people to sign up for lower property taxes, due to very high tax rates on forest land for recreation, hunting, and home sites. All the professional talk is about Sound Forestry

– Sustainable Forestry, harvesting trees that are "mature" – everything sounds good. Later - each forest owner realizes that they have to turn over their forest to an expensive outside "professional forester" and an unknown timber harvester. Then they are forced to sell their big trees for about 10% of the cost of ownership and management. And they are left with a big mess and lot of suspicion and regret.

Many of the professional foresters use the Managed Forest Law in Wisconsin to take advantage of the landowners and get them to over-harvest their timber – making the professionals lots of money at the expense of the timber grower and the natural resource they are supposed to protect and nurture.

Cover ups protect this phony system making it sustainable!

Here is the latest situation now being contended for in S Wisconsin right now – November 15, 2015:

A family has a 20 year old MFL management plan that now requires then to cut the timber off their land adjacent to their home.   This is a low value unmanaged timber land – a home site and hunting land.

Things have changed, a better plan has been proposed, but the State refused to cooperate!

No forest management was done or required to grow oak or hardwood timber – landowners were not told or encouraged how to improve their growing trees.

The State Forester 'projected' that timber would be mature at 2015 and should be clear cut.   A recently cut oak tree from the new building site shows good growth rates on the oak trees today. Most of the oak trees are not overcrowded and many trees show good form and growth. Foresters really can't predict the future.

The timber is highly variable with most areas not fitting the described mandatory harvest of oak. Black Locust and Black Cherry and Black & White Oak poletimber sized trees are showing good value growth potential at this time and would be killed for pulpwood in this mandated harvest. There are large numbers of seedling oak trees on every acre already.

No other management or harvest alternatives have been explored, encouraged, or allowed.

The landowner is being bullied to follow a plan that only benefits the foresters and logging company.

The 2015 mandatory harvest is clearly not at all about landowner's objectives—as the MFL claims.

The whole area is full of young trees

Good regeneration of a variety of species throughout . The current situation does not justify clear cutting of all trees to 2" except a few large trees said to be seed trees.

Many good quality pole size trees - cherry, locust, oak would be killed in the prime of their life—for pulp.

Prices offered are ridiculously low for the grower – way below costs of ownership. The foresters, loggers, truckers, mill owners will all make good income though the landowner is expected to sacrifice their forest for the benefit of a logging company and professional forester?

The estimate of the cutting plan says only 2% of the logs will be sawlogs and the price of $120 per MBF is very low.

98% of volume estimate is cordwood at even lower prices – outrageous low "market" prices!
The foresters tell the landowners these prices are OK and it is good to harvest timber.

Landowners are supposed to let all the wood be hauled away, then trust they will get paid fairly after the logs are delivered to distant mills. The "deal" is that 'anything is better than nothing' to justify the price paid to the landowner

Damage to the land, the few remaining trees and the regeneration will be severe from heavy machines

This large healthy white oak is growing with excellent vigor and should not be cut now.

There are much better alternatives for the forest owner, yet the DNR won't tell or teach them

What this landowner can do:

If the State requires these landowners to continue and the timber company insists on harvesting according to the signed contract, the landowners have been advised to strictly enforce the written contract provided by the timber company;

* Set up a camera to photograph and time stamp every truck entering   and leaving the property, each mill tally receipt and payment will be matched to every load

* All young tree growth bent over or held down by harvested trees must be immediately released

* All trees 2" diameter and larger must be cut/ harvested without damaging leave trees over entire area running over, pushing over small trees is not acceptable – they are to be "harvested"

* No damage to the marked trees is allowed – no machines allowed close to root area of trees

* Heavy machinery must not damage tree regeneration, soils, roads.

• All tree tops must be cut to within 36" of the ground — all of this is written in the contract!!

Landowner and their representatives will monitor the work closely and will suspend work immediately if this work is not performed properly and completed immediately

If the contract is not followed, the logging company will be responsible to pay for the damage, and no further work or wood can be hauled away until the damage is paid for in full

I believe that no logging company could possibly fulfill the contract as written and signed by their "procurement forester". If the landowner stands firm, I bet the buyer will withdraw and with no market for the wood, this forest will be saved from destruction. A much better management program should be implemented for the landowner and the community and the State.

Time For Change!

Wisconsin State Forester's actions have always been to feed the big mills with cheap wood at the expense of the landowner & the forest as well as the taxpayers.

Wisconsin DNR and Wisconsin Department of Agriculture should instead encourage local small woodworking businesses to use good local wood. Landowners should be taught to manage and value their trees – caring for trees is simple & practical and can be profitable.

Local woodworking business can perform a small annual harvest of high value wood products to benefit the community, not the decades apart slaughter for the benefit of a distant pulp corporation.

Urban Forestry Programs are failing too

Industry used to discourage utilization of city trees, but with the decline in production, that pressure has fallen off too.

The foresters use a specific disease or insect pest to create a big fuss to raise funds for sustaining their own jobs. It is silly to think that humans with a few million dollars here or there can do anything to change nature. Our best management choice is to encourage as much diversity as possible, so whatever comes along doesn't destroy a single species or forest. A healthy diverse and natural forest should be the goal of all managers, everywhere around the world. Work with nature – we will never be able to control or change natural forces.

When a disease or insect or storm kills a bunch of trees, they should be used for as many high value products right in the community - as possible. The wood chipper salesmen seem to have the greatest influence over our urban leaders - the results are clear to see.

Storm Plan – get out ahead of the Storm

Windstorms occur everyday, somewhere.   We all act like it won't happen here any time soon.

A community should have a plan to use trees blown-down in a storm to help rebuild the community.

Designate vacant lots where logs can be dropped off during a clean-up operation. A city could provide a log pick-up service

Train workers to cut trees into valuable/usable pieces and minimize damage to the logs— when possible.

Hand out a guide on log lengths, valuable parts of a tree, and what is best chipped or split.

Train workers to cut trees into logs safely and without damaging the logs (a major issue!!).

Produce a video to impress all people of the value of salvaging fallen/broken timber.

Encourage small businesses to use the wood to make high value products to benefit locals.

Provide information on kiln drying and marketing – the typical stumbling blocks to local wood.

Help rebuild the community and people's lives with the salvaged locally grown wood.

Encourage people to always choose locally grown and manufactured wood products to rebuild.

Insurance companies and FEMA should avoid cheap imported materials that export our money.

In a bad tornado, little timber value is left salvageable, but the energy value of the trees could still be used. Biofuel can be burned cleanly and efficiently and is carbon neutral – renewable. Using wood chips and pellets in a local business can pay off by keeping the values in the local economy. Using wood for fuel is in progress in other countries where big energy corporations don't control the government – people need to choose wisely and stop allowing huge corporations and lobbyists run our energy policy. A smarter approach than wasting trees needs to be implemented. Our timber is a huge renewable resource that should be used more wisely—our foresters need to do a much better job of teaching the many values of our trees!

The Even Age – Clear Cutting – Tree Planting Conspiracy

The forestry professional legitimizes clear cutting and even age management – where the industry can take all the trees at one time for their increased profit. They can even call it sustainable and certified with their magic language of smooth – sounds like "regeneration harvest".

I was trained in college that Red Oak, Douglas Fir, Yellow Pine – all were managed with even age management, as suppressed trees would not respond to the release of competition from a thinning harvest. A simple planting of new trees makes taking the old forest away supposedly OK.

One day – 30 years ago – while working in our barn workshop, I cut through a 4X4 from one of the first red oak trees I felled and had sawmilled and used the wood for our home projects. I cross cut the cant, and saw this pattern in the growth rings (above right) – it was totally shocking and changed my life in an instant of seeing a new reality. Oak trees do respond to thinning - and can be managed with selective harvesting! Since that sight, I have seen it thousands and thousands of time again – all over the world in all tree species.

When I show this piece of wood to professional foresters – 100%, every single one – close their Minds. Not one professional would look at this piece and understand and learn – they are all scared to death they will lose their cushy job if they question their training and the industry that supports their income.

There is an old fellow in Washington State – just like me – that shows that Douglas Fir can be selectively harvested and growth to old age and large size in a truly sustainable manner – crazy!! He is ignored by the professionals too.

This is just one of many phalse Phorestry practices that blindly allow the industry to take what they want for their immediate profit – at the expense of the landowner and the forest resource and the planet.

We have to stop supporting big industry and use locally grown and produced Wood and Food and Fuel!

Tree Planting – the myth

I love trees, but I don't plant trees anymore –
they all know how to regenerate without our help.
 If… We let our good trees grow and live and regenerate as naturally as possible, for as long as they are healthy and vigorous.

This is a Giant Sequoia seed— big things start small.
Our oak acorns and walnuts are quite different.

The lie of the industry is that it is OK to destroy the natural forest as long as you plant another tree... That is so stupid, we should not be faked out to think this is OK. The environmental destruction from clear cutting and most industrial logging can never be repaired by man or even by nature in our lifetime.

Planting a seedling to replace a giant is just a trick to cover the greed of the moment.

We actually have plenty of trees – and new seedlings every year – we need to see the big picture and not be fooled by simple propaganda meant to fake out the ignorant masses of people. Be Smart!

We need to stop driving off the cliff of industrialization and go back to simple methods that have supported humans for thousands of years. Maybe some things are better off in the global markets, but when we can choose something local that works to support the community and economy – we should choose that option, to balance the loss we take when we have to buy global.

Go the extra distance with local wood and food and fuel – all renewable in our communities.

The Bidding Boon Doggle

Multiple bids, tricks, fraud, cheating, insider information – all contribute to the market mayhem of the timber bidding system. Getting lump sum sealed bids is a good idea, maybe the best chance for the timber grower – but no bid in the industrial forestry system considers the real value of the wood – just offering what it takes to buy the timber from an unknowing landowner.

Contracts are often not worth the paper they are written on and often lead to false security for the forest owner. One retired State forester provided a contract to an elderly lady living in a nursing home that

basically stated– if you take those good unmarked trees, and you get caught, you will have to pay for them at the cheap price... Once this forest rape was perpetrated, no forester or government official even bothered to listen to the landowner – instead they covered up the actions of their buddy.

The timber buyer is the point of contact for the industry with forest owners – yet they go on and on taking advantage of timber growers with no regulation or control – the industry should clean this up!

Timber buyers have never paid a forest owner a price based on the value of the trees, they pay what they have to – to get the wood. This is a world-wide market now where volume is overwhelming and supply is greater than demand. If a low market price is the result of supply and demand - cheating in addition to a ridiculous price makes the situation even more absurd.

Globalization has made the timber markets worse for landowners in the USA. Our industry has a very difficult time competing with other corporations who get their wood from illegal logging operations and process it with cheap labor and few regulations that protect the environment. Even with new computerized timber harvesters (right), American companies have to be smart and lucky to survive today.

I wrote this to the local School Board. "Trees cover almost half of the entire school district – about the same as our whole State of Wisconsin.

Trees are our most abundant – yet basically unknown - agricultural crop.

The level of management of the School Forest is unfortunately the same as the surrounding forests and villages and yards. The harvest in 2013 of your pine just S of the village of Spring Green is typical and universal, yielding just a couple of dollars per acre per year income. You trusted the local government forester and a timber harvesting company from up north, and no one even knows how much you were paid or even if you were paid fairly for what was taken. This is because the market price is so ridiculously low that you just accept that something is better than nothing. For the School District to manage its own trees in this way teaches our students "industrial forestry" and perpetuates this ongoing economic disaster for all area forest and tree owners. There is an alternative you can choose now. Forest and tree owners should know how to manage this natural resource as a profitable business for the future."

For 10 years I have been trying to get the River Valley School District to manage and use their own wood from their 200 acres of School Forest. Recently I again encouraged them to build a new shelter for the school fair using the pines that are otherwise just dying every year in their woods. The engineers all said that "no engineer would approve using local wood – it has to be manufactured commercially" and only state approved engineered plans can be used today. Again, for the third time the "experts" here blocked using wood from the school forest for good local projects.

As a logger, timber buyers, sawmill owner... I don't criticize the business operations of the timber industry— This is business—a landowner must be really smart—no excuses for ignorance in this business.

Professional foresters should not accept any business practices that don't protect and nurture the forest resources and the local communities. Foresters should not force or encourage landowners to do business in this unfair and unprofessional marketplace. Ignoring their own conflicts of interest to protect their own job, salary and pension at the expense of the forest and timber growers should not be allowed. The Timber Industry and Forestry Profession are so corrupt and greedy, that as forest owners our family decided to find a better way and we did!

The oldest trick in the book – "Oops, I didn't see the fence."

Our law enforcement, our journalists, our landowners don't understand or value our trees. I recently read the Wisconsin State Journal story and heard the NBC15 TV news summary of the Rock County News of loggers caught stealing walnut trees.

This is likely the Oldest Trick in the Book and happens every single day in every single county in the region, yet it is just a tiny part of the real story on the ongoing destruction of our local forests – because no one knows anything about trees and their real value….

And usually just no one even notices.

Loggers and foresters know that they can go out and cut trees about anywhere and usually no one notices, but if they get caught taking trees, they can just pay a token payment and get away with a super bargain deal anyway – and the police don't know enough to do anything and the DNR Forestry doesn't deal with this….   It is a global problem that is rampant today even here in Southern Wisconsin.

Most trees that are cut in the world are not paid for, and even if full and fair market price is paid – the low value paid for high value trees should be a crime. Here, landowners get about 1% of the value of the finished wood products = that is like a dairy farmer selling their milk for 3 cents a gallon, a corn farmer selling a bushel for 4 cents, or a beef farmer selling meat for 5 cents a pound. No farmer would do that yet all forest owners settle for a ridiculously low market price – and in fact most are cheated in the deal.

This is unbelievable how an intelligent society like Madison Wisconsin can blindly allow our local forests and trees to be devastated for a tiny payment, our best logs are shipped to China for processing, and most of the wood products in our stores comes from another country!!

Today, there is suddenly an alternative brand of wood products coming on the market in Madison and Milwaukee that salvages and uses the thousand good trees every day that are now wasted in our cities, yards and along our streets. Wisconsin Urban Wood has developed out of the

Emerald Ash Borer scare, an now offers an alternative the Industrial Global Timber Market that continually devastates the remaining rainforests and natural forests of the planet. A group of small businesses in Madison, Spring Green, Baraboo and Milwaukee now offers wood products from the dead trees now mostly wasted in our communities.

You don't see promotion of local wood, no associations or queens or breakfasts – for a big reason -
the timber industry doesn't want to be noticed.

I believe that These are the facts today – the background story is very old and long and complicated, and things have changed and new people are now in charge. So we can simply skip ahead if if we choose, deal with it, be responsible today and just make the world a better place for our future by working together with common sense practical ideas – starting today.

I could write many books on this, these are just a few key points. Again, since the timber industry and forestry professional developed so long ago, no one alive knows the real story, so we can just skip this and be responsible humans and choose to work together now with what we have today.

We don't need to hassle about old stuff, just look at what we have today – and work together to be smart and cooperative and do what is good for the local people.

The only hope for humanity is that we choose to let the women lead – they already have the power of the two major controllers of the shopper and of sex – if they can choose to work together, their more nurturing, planning, fostering, talking manner may save us from the aggression and selfishness of men.

Buy from small business – know the grower! That is the only "certification" that can be trusted.

By eliminating all the middlemen/brokers/shippers/wholesalers… local small business can make and sell solid, natural wood products at competitive prices.

The timber industry is really good at making boards and manufactured stuff from trees around the globe, but it is not so good at managing our forests for the future and benefiting the communities that grow the trees. They are really good at propaganda and using globalization to make a few people super rich, and creating a phony Phorestry system to make people feel good about letting the planet and the people be abused.

For the timber industry and the forestry profession to continue today to use our resources and manipulate and take advantage of the people, for the profit of a few is no longer acceptable.

Is anyone else ready to stand up and do the right thing – to clean up the big act for the survival of our planet – or do we just keep playing foolish games as a few get richer and the masses suffer with dwindling resources and cheap imported stuff.

Using locally grown and manufactured wood products and biofuels is the biggest opportunity today for economic growth and ecological restoration and is a significant hope for controlling global warming.... As all products from plants are carbon neutral and energy efficient.

A new situation today

Now it is no longer OK for a rich man to have a harem, a even a mistress
it is no longer accepted for a church leader to abuse children for sex
it is no long acceptable for a tourist or businessman to have a prostitute on a trip
racism is out – spying and torture, no more – no way

Whistleblowing is suddenly cool, I hear, but not in my experience.
This guy has been blue in the face blowing the whistle on
the timber industry/forestry profession for 4 decades – if a
man blows a whistle in the forest and no one cares –
 does it make a sound?

Standing Your Ground

Our family stood up to the Sauk County zoning commission on sawmill permits a few years ago.
Government regulations are way out of control – the committee approach to regulation is crazy. Marty Krueger (county board chairman) told me face to face – "if you want to do business in Sauk County – you have to pay". The pile of laws we are supposed to obey are over six feet tall, unknown and ignored, until someone wants to hassle another. The new generation of college graduates hired as regulators and journalists and certifiers have no wisdom or experience to do their jobs, yet limited budgets mandate their employment. We are quickly losing all common sense and control . Zoning is often a self supporting department by charging fees to pay their own salary and benefits – the wrong way to govern people.

Stupid regulations block innovation and small business. Simple common sense regulation is needed that is fair and wise and treats everyone the same.   I got more pats on the back and thank yous for standing up to the county zoning than all my work to save the forests over many decades.
A profession must police itself—google the ethics of the Society of American Foresters—it is very simple

As a forester, I do believe the profession as a whole has gone way off it's ethics and responsibility.

All members must decide to be part of the flow or to stand up for ethics and work for what is right

I have tried and tried to influence the forestry profession for the better, have distanced myself, but continue to stand up for what I believe is right. I don't say or write anything until I do it myself and know what is involved. For Foresters to legitimize and participate in industrial logging and call it sustainable forestry is ridiculous.   Individuals must choose how to deal with this dilemma.

I'm a logger and Love to talk honestly with other loggers.   They are mostly just doing their jobs in a tough marketplace.
Only when they also act as the timber buyer and pay a timber grower a low price do I get angry.

This is Pete, the proud owner of a brand new Ponsse harvester (made in Finland) – working today in the River Valley School Forest near Spring Green. The protective mat was still on the floor and the clear plastic seat protector still in the cab – and it had that new harvester smell inside.

8 Wheel Drive – 330HP Mercedes Diesel Engine - State of the Art computers – 33 foot reach, Best ever, here in Wisconsin and the world.

Pete is from Northern Wisconsin – started cutting timber at age 18 – now owns his business.   His computer screen/hydraulic operated chain saw cuts timber to 1/32" tolerance.   He was cutting and sorting hardwood pulpwood, red pine pulpwood that will be hauled 100 miles to Wisconsin Rapids, sawlogs for three different sawmills up north (200 miles haul) and the best red pine were cut for utility poles for a preservative treating plant in N Wisconsin.   Most of the wood will be 2X4s and power poles, with the small stuff heading for the chippers.

None of the wood will be used around here, Pete told me.

A timber buyer from the pole plant up North marked this red pine tree with the length of pole, the red paint number showing Pete how to cut the most valuable product from this harvest.

I learned all about the current State of our State's timber industry today from a fellow logger – eager to learn what we do down here.

We had a great time sharing our experiences as he worked – from our two different worlds working in the woods of Wisconsin:

Two Different Timber Markets

Pete has the most expensive and powerful machine possible	I have a chain saw I carry by hands
Pete has to produce hundreds of cords of wood each day	I cut a tree once in a while
Pete is working 200 miles from home	I walk 100 yards down to my sawmill and shop
Pete has to work 16 hours a day to make his payments	I have no payments to the bank and work when I want to
Pete is safe and comfy and warm in his captain's chair	I work within nature in my woods
Pete never sees the wood once the trucks leaves the site	I make finished products that I install in my customers home
Pete's logs will be hauled by big trucks many times	I take my wood products to the customer in my truck when I work
Pete earns a dollar or two per tree	I earn thousands of dollars per tree
Pete is unsure of the future of this forest after he is gone	I know my forest will improve every year as I work
Pete works really hard for a living	I live for my work in my family forest

Spring Green receives little benefit from this harvest from their own School Forest
80% of the value of my salvage harvests stays right here in the local economy

I could write another whole book – but you get the gist….

Pete is the poster child for the Traditional Timber Industry today.

I know there is another way that has many benefits everyone wishes for.

Only one person knows how much timber was cut from the School Forest. No one else seems to care anyway.

After 150 years of cutting trees around here, very low prices discourage most people from getting interested in forestry.

Our family business system shows how the grower, the logger, the woodworkers, and community can all be supported by the local forest.

The Phony Phorestry Profession doesn't want anyone to know that our business system is available.
Forest Destruction Goes On As Before…

I worked as a consulting forester on commission for many years – it was normal at the start. I was the first forester to work for landowners in S Wisconsin to mark timber sales and solicit competitive bids. I always collected a 10% commission.

Initially it felt pretty good as competitive bidding quickly tripled the price paid to forest owners. The existing system was each sawmill had its territory and no one crossed the lines. The owner of the Baraboo sawmill even spoke of this at a landowner's meeting at the owner of the Sauk City sawmill told it to me straight in the face. My efforts to introduce competitive bidding got the sawmill owners of this area riled up back in 1976.

I know every aspect of marking timber and administering a competitive bidding sale. My last big job I earned $4,000 in one day marking the timber on a red oak stand. Today, foresters can earn about 5 times that in a good walnut stand with export quality veneer. Foresters use the excuse that the forest needs to be regenerated to cut down all the good trees. They work really hard to justify this ridiculous idea, but no one challenges them so the pholly continues.

On a trip to New England, I described how I worked with landowners and was paid a 10% commission for marking trees. An older forester rebuked me on the spot and told me that was an unacceptable conflict of interest. At the moment is was kind of a slap in the face, but I realized he was right. I had to change, I had to stop working on commission. I just got out of the terrible traditional timber markets that accept and rationalize all the conflicts of interest in greedy and corrupt system.

Looking back my work as a forester administering timber sales for landowners was a mix of trying to do good in a bad system. I found there was no good outcome despite the intentions or actions.

I found a better way where everything does work out for the good. Proof of this is there is great opposition from the traditional system. There is constant pressure from all sides to quit and compromise. I'm pretty fixed in staying the course now, but those around me are often swayed by the pressures to go back to the traditional big industry ways.

When I was able to triple the income of a forest owner, it was OK to get 10% of the total. Now, there is little benefit to hiring a forester – so the landowner loses that percentage for a small amount of actual work.

A forester should be on salary to do his job in a proper and ethical way. No incentive should exist to compromise the foresters jobs for short term greed. A 10% commission is a totally unacceptable and fraudulent – even Phony conflict of interest that no person should accept anywhere any how.

I've been a timber buyer many times over the years in many different situations. I change into a totally different person when I'm buying compared to when I'm selling wood and trees. The bias and conflict of interest are intense and I doubt any ethical forester can be a timber buyer in this industry and do the right thing for the forest, the forest owners, the local community, and society

Time for positive change

One of our most abundant agricultural crops that is growing all around us - is providing only 25% of the potential benefits - as our local trees are grossly mismanaged for the immediate greed of a few rich people in distant corporations – while the wood and paper products we buy in our stores are mostly imported from other countries. Buying cheap imported wood substitutes & manufactured wood products exports our money and our jobs. Billions of dollars per year vanish from the economy in the worst kind of effects of globalization.   A Phony Phorestry system works to maintain business as usual in a realm of fear of change and fear for losing their own jobs.

No one alive today was around when the timber industry and the markets for trees developed – it is a very old situation and ancient marketplace. There is no need to fuss and fight today, we can simply now see the bad situation clearly and choose to work together and do what is right for our community for our future.   For those interested in the background, and the details of the current phony workings of the phorestry profession and timber industry, that is in following chapters.

This is a win win win win thing on all levels if we just choose to get up to date with what is known and possible today.

Full Vigor Forestry and Full Value Forestry are what every forest owner and forester really wants – people need to learn there are choices to traditional industrial forestry.

Foresters don't tell other landowners about Timbergreen. Foresters Don't know or want to know, but they perpetuate the myth that our business has found a tiny niche market that only could work for me. They say the rich tourist market in Spring Green supports us and no where else in the world could duplicate our business system. Certainly not in their own local area.

Not one Forester or professor or government leader knows our business today.

From the Federal Government on down. A group of urban forestry businesses recently toured the USFS Forest Products Lab in Madison.   One of their top forestry experts told the group in his planned lesson that Solar Kilns are only appropriate for hobby use—not practical for a business. He added that they certainly don't work in the tropics and rainforests, but they do seem to work sometimes in the winter.

My book Solar Cycle Lumber Kilns was on display at the registration table in the back of the room, yet this professional forester refuses to accept that our successful kiln technology even exists. Instead he teaches old ignorant school lessons that perpetuate industrial dominance of the timber business. Our designs are being used around the planet, yet our government foresters actively work to discourage other businesses from using this amazingly successful technology.

This is disgraceful and discouraging!
Foresters work to protect their phony phorestry jobs.

I don't expect government or industry or the profession to change. But I do have hope!

Customers can support locally grown and manufactured wood, food, and biofuel products from small and medium sized local business. Simply—Don't buy from a big corporation. Look at the whole picture and the total costs to our economy and planet from buying cheap imported stuff "on sale" with "Free Shipping". Cheap imports soon clog the landfills and require rapid and repeated replacement.
The actual cost of choosing cheap imported stuff "On Sale" is very High.

Buy Grown and Made in the USA, Grown and Made in Wisconsin, Grown and Made in our community. Choose solid wood products that will last a lifetime.   We could immediately shift billions of dollars back from distant countries and into our local communities with simple choices. We still have woodworking shops and factories and sawmills that aren't being used much, that could be quickly re-energized with new sales. We are losing woodworking skills in our population quickly from under use – the sooner we boost local woodworking skills the better.
 This is a global issue.

Governments are huge buyers. Each government should choose to support local business in meaningful ways, looking at the big picture and what benefits the local taxpayers. Government spending that exports our money to buy cheap imported stuff hurts the economy.

Wisconsin State government spends millions of dollars per year to promote local food and Wisconsin Dairy, Beef, Corn, Maple Syrup etc. etc., but actually hinders and discourages the use of local wood.

Consumers should understand that whenever they buy cheap imported stuff that they are exporting our money and our jobs. If you think people should be paid a higher minimum wage, then go shopping at the big box store, or get fast food, or shop online – you are supporting the lowest wage jobs. When you buy imported stuff – you are supporting jobs that pay a few dollars per day.

Buy good value products from small business that is as local as possible. When you can buy local wood or local food, make up for the times where there is no local alternative – like electronics. Buy local doesn't mean go to the neighborhood big corporation store or shop online in your pajamas.

To grow our economy, choosing Grown and Made in the USA is an immediate and major choice we can make right now. To fix the phony Phorestry system – the profession needs to learn how to support small local business and serve the forest owners and protect the forests for our future – read on.

Landowners all need to learn that there are alternatives to what the industry and foresters have always told them. Information is totally free and available these days – there is no excuse for ignorance for anyone.

Landowners need to learn to manage their timber just like the other crops on their farms.

Timber is just like other crops, specific machines and methods are needed. Today there are many small and efficient machines to harvest trees on a small scale. Training is available to work safely and smart.

Wood is easy to process on small scale and many products can be sold direct to customers in the local community. Extra production can now be exported by taking advantage of the good tools of globalization. Way more value can be added to harvested logs, than to harvested corn, alfalfa, soy beans, wheat, etc. Tree farming should be the most profitable business of all agricultural crops. By eliminating all the middlemen/brokers/shippers – a small business can sell high quality wood products at competitive prices.

Managed Forest Law makes forestry Unprofitable

Dick Hall, writing in Agricultural Newspaper **The Country Today** on April 2, 2014 states that "Wisconsin's Managed Forest Law – a law meant to encourage Sustainable Forestry - makes forest management unprofitable even in times of stable markets. Being in MFL or Tree Farm Certification reduces or eliminates profits. A forester must be hired to write your management plan and mark your harvest. I also pay 5 percent stumpage tax on every dollar I receive from a harvest." Now he can't even lease the hunting rights on MFL lands to earn that income.

Hall also points out some aspects of the unfair traditional timber market system that a forest owner faces when selling logs or trees.

Jim Birkemeier, owner of Spring Green Timber Growers agrees. "The State of Wisconsin forestry program, including the MFL, makes forest management on private woodlots a welfare program, dependent on the government. The timber market and "sustainable" forestry system does not pay the landowner a fair price or treat them with respect. Landowners are treated like "Sheeple" by all the professional foresters – kept quiet and obedient in the flock until it is their turn for slaughter, for the benefit of the timber industry and professional foresters."

Timber prices have always been low due to an overabundance of good quality trees. That continues today and is reinforced in the global timber market. But no farmer would sell their milk for 3 cents a gallon, or their corn for 4 cents per bushel, or their beef for 5 cents a pound. So why do forest owners accept such ridiculously low prices for their trees, below their costs of production?

The timber market is a complicated and traditional institution that developed before any of us who are alive today were even born. As countries develop – trees are seen as low value and in the way of agriculture and development. Governments around the world continue to basically give standing timber to big industry to "create jobs and build the economy". No one is responsible for the awful market that forest owners face, but there is absolutely no reason we have to maintain and protect a corrupt and damaging system any longer. There is no excuse for ignorance anymore – there are better ways that could be quickly adopted to respect the growers, the land, and the forest resource.

The traditional timber industry and government forestry system have been so dominated by big powerful players, that the industry continues to cover up the scandals and silence the whistleblowers that have recently been brought into the public view, bringing needed reform and change in other global institutions.
The timber industry and government forestry systems needs to clean up its act from within and enter the modern world of ethical treatment of the resource and the people of the planet.

Local Wood Installed by Small Business—the Opposite of industry and government programs

Another forest owner called me today, asking for help to complete a "mandatory harvest" for his MFL contract. People call me several times each week, looking for a better way than the State's pressure to do a commercial timber harvest. Everyone signs up for the MFL under the guise of "Sound Forestry" and wise management, only to find they have to cut their "mature trees" at low prices and with typically high damage to their land. Landowners are forced to hire a professional forester who facilitates a commercial logging operation where the landowner feels they lose total control of their own property.

Here at Spring Green, the largest landowner is the State of Wisconsin. We together own 40,000 acres along the Wisconsin Riverway, and there are many parks and State Forests. Several State timber harvests at Spring Green have resulted in big harvests that were sold to companies up North for small amounts of money, with little benefit to our community at all.

The most recent cutting just South of Spring Green in 2013 was administered by the DNR Riverway forester based at Tower Hill State Park. Two tracts were harvested – one State owned pine plantation and a 12 acre pine stand owned by the River Valley School Forest. While the State land was a first cut, the School Forest had been thinned several times and the trees pruned to increase their value. The negotiated price was the same for both sales, $12 per ton. This may be market price for pulpwood, but the School Forest was not given credit for the sawbolts and larger sawlogs that were sorted out and trucked away as higher value products. No one verified the actual amount of wood taken. Again, the trees were sold to a distant company with little benefit staying in the community. The income for the State and the School was about $3 per acre per year – not even beginning to cover the costs of owning land and managing timber. The foresters and loggers all made good money though!

A recent hardwood harvest on State Land created little benefits for the local community

For the State to set this type of precedent hurts all landowners and teaches everyone that timber is not valuable or manageable in a profitable way.

Over the past few decades, soaring property tax rates on forest land for "recreational hunting" or home site potential has driven tens of thousands of landowners to accept the promise of "sound forestry" to manage their land at reduced property tax rates. The primary required action in a MFL management plan is to harvest "mature timber". Many landowners who trust that "sound forestry" will be a good thing, find out too late it really means they are forced to cut their good big trees at ridiculously low prices with typical high damage to their land.

The State of Wisconsin forces landowners to harvest wood at market prices that don't even cover the costs of owning land and managing the forest. Timber growers are not paid a reasonable or profitable value for their trees, yet everyone else in the forestry profession and logging industry earn a profitable income – and many get really rich of the controlled ignorance of the landowner.

The MFL dictates that forest owners are not qualified to manage their own land and timber crop.

Landowners are required to pay a "certified" management plan writer and hire a "professional" forester to mark the trees for harvest. Foresters routinely tell the landowner they are not smart enough or able to manage their own forester, only a professional can do that.

The professional foresters that landowners are required to hire, generally still are paid on a sales commission, an unethical practice that is fostered by the profession. The worst abuses I have seen come when a retired DNR forester goes into business as a private forester and uses their inside knowledge to abuse the system for their own greed. Every time I try to expose these incidents of landowner and forest rape, the DNR foresters in Madison cover up the allegations against their buddies and blame the landowner.

I tried every program of the government forestry system and our family harvested timber in the traditional timber markets – all of it is bad for a caring and informed and involved landowner. I quit!!

I found a better way and learned to do just the opposite of what the timber industry and forestry profession wanted us to do, then found it was exactly what every forest owner and forest manager should actually want!

I developed a new and separate market for our trees where the landowner is in control and uses the best of the best ideas from around the world to the advantage of the local community, not a big distant corporation. We salvage the dead trees—using what Nature gives us—letting the good trees grow!

Our forest makes a good income. Our forest supports local jobs and the community.

Hall's conclusions are State of the Art – in the traditional timber industry, but still in the old school mentality that something is better than nothing. A **1 dollar profit** is not sustainable or fair or just.

True sustainability is where the grower/producer earns enough to have extra income to reinvest in their business and improve each year and have extra to share with others and have even more to work to improve the future of the planet earth. No one executive gets rich and takes advantage of the masses.

Globalization will either save us or kill us people all off – nature and the Earth will continue, no problem.

It is our choice – there is no excuse for ignorance today and tomorrow.

Another Very Experienced Timber Growers Comments

My last timber sale was also a disappointment ----I had some fine veneer trees in the 20-26 inch category ----yet the price did not reflect the value of 46 years of forest management.

A harvest is subject to the whims of the logger and the lowest common denominator price for trees that will not compensate for the time, money, risk, and opportunity cost of the landowner to raise quality trees.   There is not enough money generated in a timber sale of average or better than average timber for the small landowner.

The time and effort the small landowner has to put into his property to properly manage for quality timber is not worth the effort in terms of income generated. Value received for the effort at present makes an alternative investment in almost anything a better value.

The old axiom of "cut and run" is true for all timber companies I have dealt with for 46 years, cutting more than 400,000 board feet of timber in numerous sales of hardwood.. In spite of a bid process and a signed contract done by a professional forester as recommended in todays forestry practice - the mess left behind in terms of damage to standing and young timber trees is extensive. Getting them to live up to the contract is difficult at best----- especially after they have loaded the cut trees and left the property.

(Prices for good Red Oak have fallen—Walnut prices are strong today due to the export market.)

Timber companies only want the best trees. As you are very much aware trying to keep up with "best forestry practices" is tough -----the cost of doing so has escalated far faster than the price of logs and timber.

The last two sales of walnut from our place have gone directly to Dubuque; loaded on a barge, down the Mississippi to New Orleans and loaded on a freighter for the Far East---not just China. It is a crime that the last and only person who actually touched each log was the logger who cut it down. No saw mill, no kiln drying, no marketing of the wood and no ultimate product development, marketing, and sales in this country. But what else is new---we have exported just about every industry in this country to the Far East and now other Third World countries.

I have yet to find a logger who gives a damn about the forest remaining after a cut and run logging. To be an isolated small land owner and raising high quality trees for lumber and veneer does not make economic sense for younger people.

You should not use my name or quote me directly----timber companies will treat me as a pariah and I may not be able to sell any wood at any price. (he is also enslaved to the DNR programs) (Most forest owners are afraid too)

What Forestry Should Be—What we do at Timbergreen Farm

We make the Most Stainable Products in Wisconsin –
according to the Wisconsin Sustainable Business Council

At Spring Green Timber Growers – we keep the many value of forests and trees in the local economy

Timber Growers masters the Native North American teaching of using just the dead and dying trees   that the forest gives you each year - so the trees will last forever.
Never allow industrial demand to determine what trees are taken.

Timber Growers encourages natural succession and natural regeneration,
following the German Dauerwald teaching - "Watch Nature".

Timber Growers carefully harvests an average of one tree per acre each year
so the forest is never significantly changed or damaged.

Timber Growers practices Arthroscopic Logging to improve the forest
using the smallest equipment possible,
doing the least damage and most good.

Timber Growers utilizes each part of the tree for its highest value use!

Timber Growers earns a minimum of $10,000 per thousand board feet
$10 per board foot pays a good wage and covers the costs, plus profit.

The opportunity at this income level is to create one good job for every 10 acres of forest in this region, and for every 50 trees cut in one year in an urban forest.
Most of our products earn a much higher return per board foot of wood.

Timber Growers uses free natural wind power and solar heat to dry our lumber
using Timbergreen Farm's unique Solar Cycle lumber dry kilns.

The nightly moisture equalization period is the key to success
that produces superior quality lumber.

Timber Growers makes hundreds of different high-value finished products,
putting local people to work using salvaged dead and dying trees.

Timber Growers sells wood products direct to customers,
usually earning full retail prices,
keeping nearly all of the money in the local economy.

Timber Growers installs and finishes mixed species custom blended wood flooring,
cabinets, stairways, furniture, etc. right in their customers' home.

Timber Growers shares all of their information with other landowners to encourage them to earn good benefits too.

**Using locally grown and manufactured forest products directly lowers
the demand to clear cut the remaining natural rainforests in the tropics.**

Use **LOCAL** Wood
It is **LOGICAL**
a **G**ood **I**dea

Most of the wood products on the market today are from trees that were illegally logged or obtained by some measure of fraud, then processed by the cheapest labor working in poor conditions. Environmental protection is neglected in most all steps of the manufacturing to produce the cheapest possible stuff for the huge profits of the big corporations.

Just a question of time—does any body care?

An unexpected meeting with an 80 year old forest owner my first day in Santarem Brazil, immediately clarified why I was right there. He told me,

"Our State of Para, Brazil, is the number one center of forest devastation in Brazil. The Timber Business is an 'Industry of Deforestation', they are hacking away at huge areas of the rainforest every day. The 3 big sawmills south of our city are getting most of their logs from harvests that are cutting down the primary rainforest. These sawmills are exporting the logs and lumber. The logging slash is burned and the land turned over to the cattle ranchers or soybean farmers. Today there is a lot of corruption and illegal logging, despite all of the International pressure and work by hundreds of organizations.

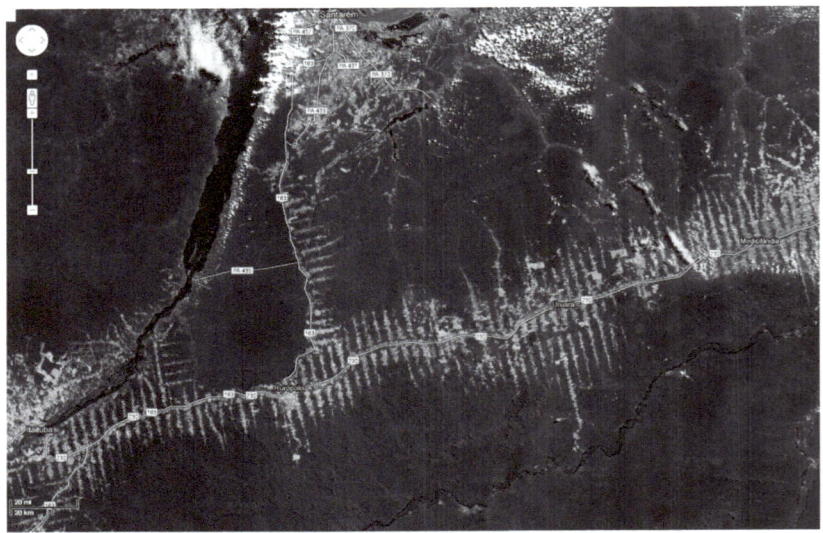

A satellite image of the Brazilian Rainforest being clear-cut along a new highway just South of Santarem. Industry is eating away more every day.

The forest doesn't stand a chance - **It is just a question of time, Brazil will be just like the USA - with no significant Old Growth Timber left."**

Forest owners in Brazil see this ongoing deforestation everyday but feel hopeless to do anything. In the villages along the great rivers, everyone knows the illegal logging and political corruption is sustained by the global markets – and the aid programs from governments and non-profit organizations are sustained by the rich nations grabbing their dwindling natural resources.

The use of Trophy Wood should be shunned and shamed just like the taking of trophy animals

For the forestry profession to allow and accept and participate in the ongoing destruction of our remaining natural forests by industrial logging - and even call it Sustainable Forestry is disgraceful.

    I couldn't participate after 4 years working as a forester. I got out—and worked out a much better timber market from the forest owner's point of view.

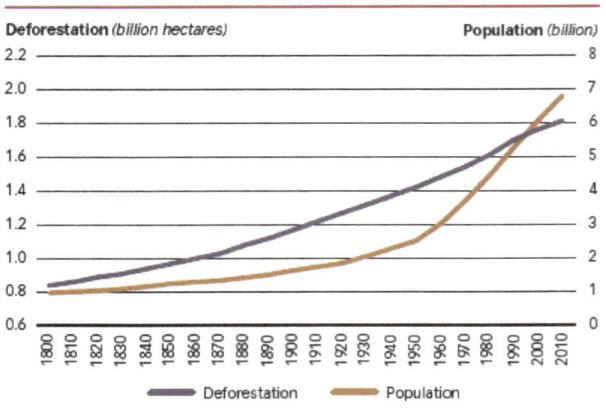

Figure 1: World population and cumulative deforestation, 1800 to 2010

Selling and buying timber and getting paid on commission are all extreme conflicts of interest for any forester. Many foresters are exposed to this practice as normal and acceptable—the forestry profession must immediately purge these practices from their membership. Foresters need to teach then follow the ethics required by the profession of forestry.

Foresters are active participants of the exporting of our best logs and lumber right now, in the feeding frenzy for the Walnuts growing here.

Wisconsin DNR Foresters should promote Good Local Wood Grown and Made in Wisconsin.

Foresters should teach common sense management practices so landowners can care for their trees, not tell the forest owners that they are not qualified to manage their timber.

Foresters should lead by example, not just talk about Sustainable Forestry.

Foresters should encourage landowners and communities to value and care for their trees.

These are some of my ideas after 40 years in the timber industry and forestry profession.

The high value of finished wood products should reward and support the local community, not just a few greedy and powerful people.

I don't expect foresters to suddenly switch to Full Value Forestry. I hope that foresters will take the time to actually learn what we do before they talk about it, or decide. I hope that foresters will realize that Timber Growers can manage their own forest and market their own timber, and allow that option to be explored, taught, and implemented. Teach people everywhere all of the options that are out

there, and the real costs and benefits of alternative markets and methods.

If anyone cares, a lot more can be discussed

Jim Birkemeier—forest owner
Spring Green Timber Growers

About the Author

Jim Birkemeier has been a strong advocate for the region's forests and landowners for 40 years. His family bought timberland 47 years ago and immediately began planting trees and caring for hardwood forests. After graduation from University of Wisconsin – Madison in 1976 with a BS in Forest Science, Birkemeier immediately went into private forestry business to help landowners earn more income from their timber. Within weeks, he began to stand up against the typical "rip-offs" faced by inexperienced landowners dealing with industrial timber buyers. After a few years working as a forester in the timber industry in Wisconsin, Birkemeier quit in disgust and discouragement due to the poor treatment of the forest and forest owner by the industry, government forestry programs, the educational system, and the forestry profession as a whole. Working on the family farm, he re-learned forest management, from the landowner's point of view. His gut notion became the guiding light – **Do Just The Opposite** of what the professionals tell the landowner to do. **It has worked out perfectly for 3 decades.**

Now Birkemeier lets all the good trees grow as long as they are healthy and vigorous.
He quickly learned to use all the left-over trees that industry considered waste. Mixed Species Custom Blended hardwood flooring, cutting boards, and wooden ornaments now earn the same high value for all species that grow on the farm. Methods were learned to safely harvest and then manufacture high value wood products from the many thousands of crooked and bent over trees left from previous logging where only the good straight red oak trees were taken. And he developed high value products from small diameter logs that the experts teach are just pulpwood or cordwood or firewood.

Birkemeier always has struggled with the government foresters to try and make growing and managing timber a profitable business. The Wisconsin DNR Foresters soon developed a rubber stamp and labeled all of his work, management plan with: "This work does not meet government specifications!" Anything with Birkemeier on it was rejected.

Birkemeier's training programs for landowners at Timbergreen Farm developed into the Sustainable Woods Cooperative movement in 1997. A group of landowners put together their ideas for running a successful community based forest management and wood products business. The cooperative sprung to life until the people were mostly wrangled back into obedient flocks of Sheeple by foresters with their State badges, free money, and influence.

While local experts shunned his innovations and efforts, his wood manufacturing and marketing system did catch the attention of the United Nations FAO and other organizations. Birkemeier has presented his ideas at 4 International Conferences on Forestry in Viet Nam, India, Thailand, and New Zealand. Birkemeier has taught his unique system to other landowners in 20 countries and hosted forest owners from 80 countries to tour Timbergreen Farm. Sharing information with other timber growers is the focus of his work today.

The author of three books; Full Vigor Forestry, Full Value Forestry, and Solar Cycle Lumber Dry Kilns as well as a full series of DVD presentations - and reportedly the most extensive forestry website on the internet – he shares everything he knows and does.

His business was just awarded The Most Sustainable Products in Wisconsin by the Wisconsin Sustainable Business Council. Last year it was the award for the Best Sawmill Business on the continent by WoodMizer. Best Small Family Business in Wisconsin and the Market Spark Award in the Madison Independent Business Awards also recognized the family business.